John Flowerdew

Woodmice
and Yellow-necked Mice

Anthony Nelson

© 1984 The Mammal Society
First published in 1984 by Anthony Nelson Ltd
PO Box 9, Oswestry, Shropshire SY11 1BY, England.

All rights reserved. No part of this book may be reproduced, stored in a retrieval system, or transmitted in any form or by any means, electronic, mechanical, photocopying or otherwise, without the permission of the publisher.

Series editor Robert Burton
Drawings by Jean Vaughan (with permission of the Zoological Society of London) and Graham Allen
Photographs by courtesy of Jane Burton, Robert Burton, John Flowerdew, James Mair, Pat Morris, Claire Nuttall and Kim Taylor

Royalties from this series will go to the Mammal Society

British Library Cataloguing in Publication Data

Flowerdew, John
Wood mice and yellow necked mice.—
(Mammal Society booklets)
1. Mice—Juvenile literature
I. Title II. Series
599.32'33   QL737.R6

ISBN 0 904614 13 1

Designed by Alan Bartram
Printed by Livesey Ltd, 7 St John's Hill, Shrewsbury

The wood mouse, also called the long-tailed field mouse, is the most common wild rodent in Britain. An inhabitant mainly of woods and fields, it is also found in gardens, outbuildings and houses. At first glance it may be confused with the house mouse, but there are many features which distinguish the two species. The habit of leaping high into the air when disturbed is a good clue to the wood mouse, as are the prominent eyes and ears which give it a particularly appealing look, and the tail which is as long as the head and body.

The wood mouse *Apodemus sylvaticus* is even more likely to be confused with the yellow-necked mouse *Apodemus flavicollis* which is very similar in appearance and habits. The differences between the two mice are described on page 22. Both the wood mouse and the yellow-necked mouse belong to the family Muridae, to which other mice and rats belong.

Like other small mammals, wood mice are not often seen, unless their hiding places are disturbed and they run for cover. Much of our knowledge of their lives has come from trapping. By using live traps, the mice can be examined and released unharmed. Before being let go, they are marked so that they can be identified when recaptured at a later date, to give information on their movements and life history. They are also examined and weighed to provide details of sex and age. The traps are put out in lines or grids at periods throughout the year so that a picture is built up of the lives of individual mice and the changes in the population. Comparing populations shows how the mice react to living in various habitats, with different food supplies and predators. The changes in the size and structure of the populations helps to

Fig 1. Sequence of wear on the molar teeth (after Delaney and Davis).

show how the population numbers of small mammals are limited or regulated under natural conditions.

In the following pages, information is given on how wood mice are studied – by measurements, examination of sex and reproductive state and so on – and the results of population studies are described. However, despite all this work on a very common animal, there is still plenty to be learned, especially about the social life of the species.

Description

As wood mice are regularly caught by naturalists studying small mammals, or are brought in by cats, a full description is given to help distinguish them from house mice, and also to give clues to age and sex of the specimens. Juvenile wood mice, up to about seven or eight weeks of age, have a dull greyish-brown fur with a pale belly. They moult, when five to eight weeks old, into the adult dark brown fur on the back, which becomes more yellow towards the flanks. The adults also moult, often between August and January but signs of moulting can be found at any time of the year.

The house mouse, by comparison, is more grey-brown in colour and the fur seems shorter with no distinct line marking the change from the upper fur to the paler underside. The tail of the wood mouse is dark brown above and pale below and is lightly haired, whereas the tail of the house mouse has a more scaly appearance with no colour difference, and is slightly thicker towards the base than that of the wood mouse.

Skulls often turn up in owl pellets or in the remains of carnivore meals. When examining a skull look particularly at the teeth. These have the typical rodent pattern of a pair of incisors (the protruding front teeth), which grow continuously from the root to replace the tips worn down by gnawing,

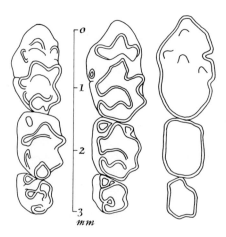

Fig 2. Pattern of cusps on the unworn molars (after Clevedon Brown and Twigg).

separated by a gap (the diastema) from the molars or cheek teeth, which are used for chewing. The teeth of the wood mouse have a concave wearing surface on the upper incisors, in contrast to the notched surface seen in the house mouse. The first upper molar (the first cusped tooth in the upper jaw) has four roots with corresponding alveoli (sockets) in the skull, in comparison with the five roots of the harvest mouse and rats and the three roots of the house mouse. The upper incisors generally measure 1.1–1.3 millimetres from front to back across the unworn upper part whereas those of the yellow-necked mouse measure 1.45–1.65 millimetres.

When measurements fall between the two, the skulls may be classified as 'old wood mice' or 'young yellow-necked mice' by the amount of wear on the molar teeth. The wear on the cusps of the molars increases with age so that they are worn down in the sequence shown in Figure 1, and in later life they form a flat grinding surface. The stage in the sequence of cusp wear may be used to classify mice into age groups, but the amount of wear depends on the types of food being eaten. Thus it is not possible to give an accurate age to the stages, unless a reference collection has been made by collecting skulls of mice of known age from the same habitat.

There is little difference between the sexes in skull measurements except for some island populations, but there are marked sexual differences in the shape and size of the paired *os coxae* which form the pelvis (Figure 3). These differences occur after the onset of puberty. In the female the bone structure changes during pregnancy to make birth easier. The study of rodent *os coxae* is valuable in owl pellet analysis because they can be used to identify remains to genus and often to species and sex.

The external features of the sexes differ only slightly, but there is a very noticeable dark scrotal sac protruding under the base of the tail in adult males. Females show bare patches around the prominent nipples during lactation.

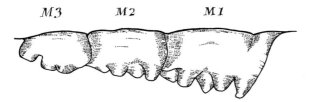

The vaginal orifice, close to the anus, is closed (the imperforate condition) until puberty. The open (perforate) orifice will close again during mid-pregnancy and in winter anoestrus, opening for parturition and post-partum oestrus or the start of the breeding season in spring. Immature females may show a slight indentation of the skin over the vagina which marks the place where perforation will occur at puberty, and the covering in mid-pregnancy or in winter sometimes has a scaly appearance.

Fig 3. Identification of rodents by the os coxa of the pelvis. (a) Male wood mouse with straight posterior margin and (b) female with concave margin. (c) Male house mouse with convex margin. (d) Female house mouse is less concave than the female wood mouse. (e) Bank vole has narrow, waisted and rounded pubis (anterior projection).

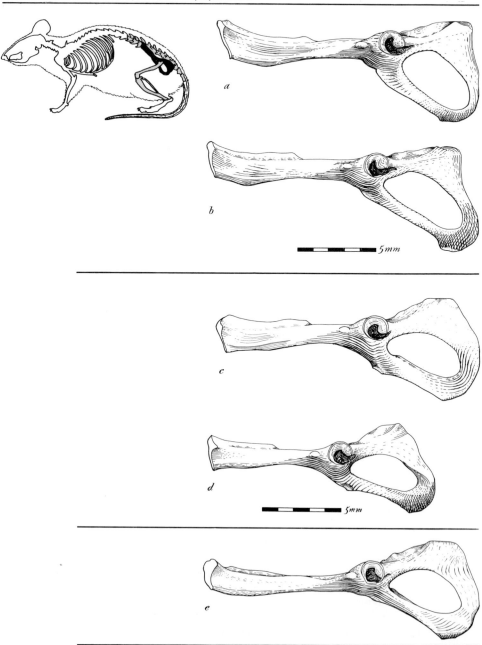

Measurements

The length of the head and body usually falls within the range of 80–100 millimetres in adults but some reach 110 millimetres. Body weights may average 16–18 grams in winter and 25–27 grams in summer but 37 grams may be reached by old males and heavily pregnant females. The length of the hind foot is 20–23 millimetres. Island populations attain much larger measurements; the largest male recorded from a study of wood mice on St Kilda (Outer Hebrides) was 54.2 grams and the largest female 48.5 grams.

Distribution

The wood mouse is distributed throughout Europe, including Iceland but not northern Scandinavia, and extends eastward across central Asia to the Altai and Himalayan mountains. It reaches south to the northern parts of Arabia and North Africa and is present on most Mediterranean islands. It is generally an inhabitant of the deciduous woodland and steppe zones and extends only a short way northwards into the coniferous zone.

In Britain and Ireland the wood mouse is found in most habitats except in open mountainous areas and where there is much grazed grassland. Occupied British islands include the Isle of Man, Isle of Wight, Skomer, Bardsey, most west Scottish islands (including St Kilda), Orkney, Shetland (including Fair Isle and Foula), all the Channel Islands, and Tresco and St Mary's in the Scilly Isles. Most of the smaller islands have records of the species but it is apparently absent from Lundy, Lunga, North Rona and the Isle of May. Irish islands with records of wood mice include Lambay, Clare, Inishmore, North Bull, Rathlin, Achill, Bere, Inishboffin, Sherkin, Valencia, Tory, Great Blasket and Cape Clear.

Wood mice in history

Differences in the skulls suggest that there are two distinct groups of wood mice on the mainland of Britain. There is no difference in size and colour between the two groups. One group lives in the western and central part of the country and the other in the east. It is possible that they evolved during the Ice Ages in different areas not covered by glaciers and later spread out to colonise their present ranges.

Of the island populations, wood mice on Jersey, Guernsey and St Mary's (Scilly Isles) are similar to those on the mainland and they are possibly descendants of Ice Age survivors. The populations on Alderney, Sark, Herm (Channel Islands) and on Tresco (Scilly Isles) differ considerably from their neighbours and are possibly descended from mice brought in by man or have diverged *in situ*. In the Scilly Isles there is a further possibility that the two present populations might be the result of separate introductions from different sources. The mice could have been brought in quite easily with sacks

Fig 4. Distribution of yellow-necked mouse. (Prepared by the Biological Records Centre from Mammal Society data.)

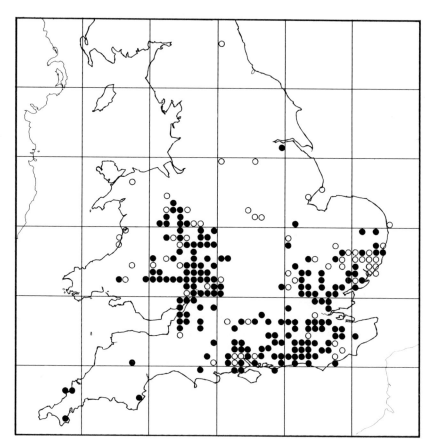

of grain or other food in the same way as rats and mice have been taken around the world in recent times.

It is thought that no wood mice survived the Ice Ages in the Hebrides and Shetlands (except perhaps those on St Kilda) so that they, too, must have been introduced by man. The Vikings are the most likely candidates to have carried the mice because present-day wood mice on these islands show affinities with Norwegian wood mice. Irish wood mice are also thought to have been introduced by man.

Variation

At one time some Scottish islands were believed to have their own individual subspecies of wood mice, but it is now generally accepted that these are no more than variants of the mainland form. They had been based on colour and size, which are now known to be factors that overlap widely between

Figs 5 and 6. Who ate that?
Left: Empty hazelnut shells, showing teethmarks.
Right: Ash seeds.

populations. Indeed, many island forms have been kept together in the laboratory and have interbred. In general, however, island mice are larger than those from the mainland (see *Measurements*).

A white tail-tip is very common in some populations (*eg* Tory Island, Ireland) but a sample from England showed only 1.9 per cent with white tail tips and another from Scotland 4 per cent. There are records of piebald, albino, yellow-grey (called isabelline), silver-grey and melanic varieties, the last form being recently recorded from East Anglia and Hampshire.

Signs

The droppings are usually dark brown or black and of variable length and width. They are often longer and thicker and have more rounded ends than those of the house mouse and are often larger and darker than those of the

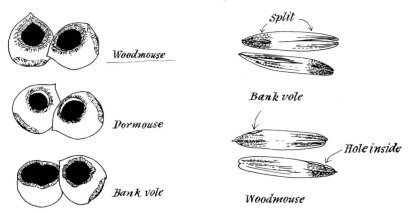

bank vole. A typical length measurement would be 2.5 millimetres for an adult. Collections of droppings are difficult to find in the wild and they appear to be scattered at random. However, in buildings they are left near feeding sites and sometimes piled in corners and near the nest.

Signs of feeding are common in all habitats (Figures 5 and 6). Disused birds' nests, particularly those in hedges, are often filled with the remains of past mouse or vole meals. Wood mice tend to leave the flesh of hips, haws and other fruit and eat only the pips, while voles eat the flesh and discard the pips. On the woodland floor fallen tree trunks or stumps may be used in a similar way and strewn with food remains, but more often the food is taken to the shelter of the base of a tree or under a ledge.

Tracks in the snow show that wood mice often change from normal walking to move by jumping. The four footprints are close together and sometimes partially overlapping, with a distinct difference in size between the fore and hind prints. The tail usually leaves a long mark behind the tracks.

Fig 7. Track of wood mouse.

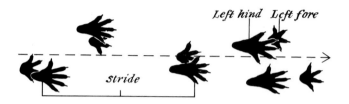

Habitat

Wood mice are found anywhere from flooded sedge fens to inhabited buildings but they are typically present in woodlands, hedgerows and arable land. In woodland and hedgerows wood mice and bank voles live side by side, but wood mice are usually the only inhabitants of cereal fields, root crops and other arable land, unless house mice accompany them. They remain on fields even after ploughing and stubble-burning, although numbers may be reduced for a while. In woodland there is usually no particular preference for areas of good cover, and wood mice will venture into the middle of bare areas of leaf litter or earth. The presence of wood mice in coniferous woodland may vary with the food supply available; moderate densities of 30 mice per hectare have been found in mature larch plantations where there are abundant prepupae of the larch sawfly in the ground. However, population density in conifer woods may be much lower, perhaps less than 10 mice per hectare. Wood mice are common in gardens and sometimes in grass and heather, especially if short-tailed voles, the dominant grassland rodents, are absent. In Ireland, despite the lack of short-tailed voles, more captures of wood mice have been made in mixed heather, bracken and grass than in areas of grass only, which suggests that wood mice do not favour pure grassland.

Stone walls form excellent homes for wood mice and, if the adjoining fields are grazed, the mice do not move far from the wall. On islands, the preferred habitats are often walls, cliff-tops and rocky outcrops. On St Kilda Island wood mice are still thriving whereas house mice became extinct after the human population was evacuated in 1930. Although house mice do survive on other deserted islands it seems likely that they cannot compete with the wood mice in the absence of man.

Behaviour

Wood mice run with a fast, darting motion, and often jump when disturbed. When they are searching for food or exploring, they walk with the head down and the tail straight behind them, sometimes taking slow, deliberate steps. Occasionally they will stand on their hind legs, using the tail as an extra support, and point their nose upwards as if investigating a new scent. Grooming is meticulous; the face and ears are wiped with the forefeet and the tail is drawn through the mouth. The flanks and underside are groomed by the mouth, while the hind feet cover the neck, sides and underside as well.

Wood mice can find the entrances to their underground runways with remarkable speed even in dense cover, and it has recently been shown that they can orientate themselves using the earth's magnetic field. They are probably using their senses of smell and vision as well, and they will also have a good memory of the exact location of the entrances. They are almost wholly nocturnal, coming out of the burrow at dusk and usually continuing to be active, with perhaps short rests, until dawn. Males are usually more active than females.

During long winter nights there are two peaks of activity, one a few hours after dusk and the second a few hours before dawn, either side of a longer stay in the nest in the middle of the night. In summer there is only one peak of activity, and some mice stay out all night. Wood mice are seldom seen by day, but there may be some activity underground, and breeding females come out briefly around midday during summer days in Scotland, where the nights are short. Bright moonlit nights inhibit activity and cold conditions with heavy rain also keep the mice in their nests. Warm wet nights, however, often produce good catches in traps.

Home range

The size of the home range varies between sexes, seasons, habitats, and methods of study and analysis. Males usually have larger home ranges than females and winter movements are smaller than those made in summer. Trapping and tracking studies in woodland indicate that average home ranges for males vary from 0.18 to 0.31 hectares and that those of females vary from an average of 0.01 to 0.21 hectares, but the ranges of a few individuals may reach 2 hectares or more. A radio-tracking study of wood mice in woodland, using the same method for estimating the size of the home range as in the trapping studies, has given average home range areas for males of 1.1 hectares and 0.4 hectares for females.

As the juvenile male becomes sexually mature he will slowly increase his range so that within two months of leaving the nest he will have a range nearly as large as those of older adults. However a young male born towards the end of the breeding season, from September onwards, does not attain sexual maturity immediately. His home range increases only moderately after two months and is likely to stay small through the winter until the start of the breeding season in February or March. The home ranges of these old males which have survived the winter are usually larger than those of males which were born after the winter.

Only a small amount of dispersal movement (long distance movement away from the home range) has been detected by trapping studies within woodlands. However, movements out of woodland in spring and a return in autumn are common where suitable habitats such as cereal fields are close-by; the return may be associated with harvesting.

There is no hibernation but a wood mouse is occasionally found in a state of hypothermia (reduced body temperature) in cold weather. It is not known how common this behaviour is, but it will help the mouse to save energy. Communal nesting in winter is known to reduce energy demands and the quantity of material in the nest in laboratory cages is increased under cold conditions to improve insulation.

Burrows and runs

Most wood mice live in underground burrow systems. They are excavated at a depth of 70–180 millimetres, but may go deeper if they connect with a mole run. The diameter of the burrow is 30–40 millimetres, becoming wider at nest chambers and food stores. The burrows are presumably occupied by generation after generation of mice and are enlarged and modified as required. The tunnel often leads into the decaying roots of trees or into tree stumps, under which nest chambers and food stores are often found. The burrows are linked by a series of runways over the ground, or above ground level via fallen branches or tree trunks and similar obstacles. In rocky areas, they pass within cracks and tunnels formed between rocks or scree. It is not unusual for nests to be made above ground level in bird nestboxes, and mice commonly forage above ground level.

The entrances to burrows are sometimes blocked by leaves, twigs or stones, and even Longworth live-traps may be similarly obstructed, particularly in autumn and winter. This behaviour probably serves two functions: to disguise the entrance to the burrow, nest chamber and food store, and to aid storing food. Wood mice are compulsive blockers-up of holes and storers of food in winter and they sometimes construct small piles or 'cairns' of stones or debris at the entrances to burrows or nearby. One East Anglian farmer found his field drains blocked up for hundreds of metres by acorns carefully carried by wood mice from a nearby oak tree! Forgotten stores of seeds and fruits are responsible for the successful regeneration of many trees and shrubs.

Burrow systems are commonly found in ploughed and cultivated fields, and the entrances to newly-excavated burrows have a mound of earth 'spoil' outside, often in the shape of a flat cone. This earth soon blends in with the surroundings and is lost. If burrows are destroyed, as the shallower ones will be during ploughing, then new ones are quickly excavated. A single mouse can move 1–3 kilograms of earth in two hours. In direct-drilled (seeded without ploughing) cereal fields, the entrances to the burrows are joined by obvious well-worn tracks on the soil surface. An examination of two burrows in a winter wheat field showed that they were curved tunnels 250 millimetres and 150 millimetres long ending 65 millimetres below the surface. Several mice may emerge from a single burrow system when forced out by ploughing.

The nest is made of whatever is locally available. One nest in a garage was made of leaves, pieces of onion, wood chippings and old rags; another in a

Fig 8. Burrow system of a wood mouse (after Jennings).

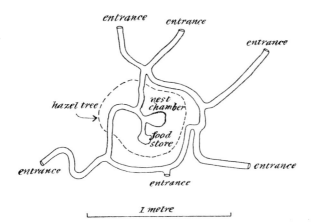

potato harvester standing near a hedge contained many dead leaves and strands of grass. Although the nests are commonly found in the burrow systems they may also occur in grass tussocks, hedgerows, holes in trees, bird nest boxes and buildings.

Social structure

The social structure of wood mice changes from winter to summer, in that individuals may nest communally in winter but not apparently in summer. It is thought that associations of a single sex may be formed in winter when the home ranges of individuals overlap to a great extent.

In the spring the breeding females take up mutually exclusive home ranges and nest singly. This lack of overlap of home range of females, in contrast with the large overlaps seen between adult male home ranges and between males and females, suggests that females are territorial and defend at least part of their home range, and that adult males compete for access to breeding females. Whether this interpretation is true for wood mice in all habitats and at different population densities remains to be seen. Observations at a baiting point have shown that there is a dominance hierarchy in feeding. Only when one pair of mice moved away from the food would others move in. It has been suggested that the breeding pair may possibly become a 'superfamily' with a number of females and subordinate males dominated by a male, but the true social relationships between males and females during the breeding season need further study.

Individual wood mice can become highly aggressive, and in the laboratory adult males will kill juveniles intruding into their cage. Encounters between breeding females and juveniles can also be very aggressive. However, most encounters are passive and in the laboratory environment of a neutral cage, aggressive behaviour is uncommon; in the wild, wood mice are most likely to avoid each other. Adult males are usually dominant to females and juveniles and the subordinates usually show submissive behaviour, or are just passive and keep away.

Communication

Communication is apparently visual and tactile at close-quarters, but the role of sound and scent signals is not clear. Ultrasounds, too high-pitched for the human ear, are produced by the young up to 24 days old and are induced by exposing them to low temperatures. They cause the mother to pick them up and carry them back to the nest. Squeaks audible to the human ear and ultrasounds are emitted by males when sniffing females, during attempted and successful mating, and after disturbances to the nest.

The scents from urine, body secretions and from subcaudal glands on the lower surface of the tail are probably important in communicating the identity of an individual to its neighbours. The large size of the subcaudal sebaceous glandular area makes the tail of adult males thicker at the base than that of females and juveniles, and the milky secretion which can be squeezed from it differs chemically between the sexes, between adults and juveniles and between populations in different places. The chemical information in the secretion may be used as olfactory signals deposited on the ground or in burrows. Whether the secretion is deliberately rubbed on the ground or is rubbed off when the individual is sitting or stretching is uncertain; no deliberate marking behaviour has been observed.

Food

In woodland, seeds are the main food, particularly in the autumn and winter, but seedlings, buds, fruits, nuts, snails, arthropods (insects, spiders, woodlice and allies), fungi, moss and bark are also taken. The preference for seed food gives way to insects if they are abundant, as when winter moth and other caterpillars fall from oak trees to pupate in the soil in May or June or when sawfly prepupae are abundant in larch woods (see *Habitats*). Other common insect foods are beetle adults and larvae, fly larvae and pupae, and caterpillars in general. If neither seed nor insect food is readily available, then buds and shoots are eaten; this often happens in woodland in March or April.

On arable land the diet is influenced by the present and past crops, and it is likely to vary according to the weeds and arthropods available as well. In fields of winter wheat and land ploughed after the harvest, sown and shed grain is taken in quantity in the autumn and early winter. Arthropods continue to be taken into early summer when the bulk of the diet is made up of weed seeds and grass flowers. On sugar beet fields, before harvesting in autumn, weed seeds form a large part of the diet. Larval and adult beetles and larval butterflies, moths and flies form part of the diet in winter. In late winter the

Woodmouse forage at night.

Food
Seeds, fruits, nuts, snails, insects, spiders, woodlice, caterpillars, larvae, bulbs, peas, beans, tomatoes

amount of cereal seed and weed seed taken declines, arthropods continue to be taken and earthworms are added to the diet as well as remains of sugar beet root if it is available. In potato fields which have been harvested in autumn, the signs of feeding by wood mice can be seen for much of the winter in the remaining potatoes.

In gardens wood mice are notorious for taking bulbs, beans, peas and tomatoes, and many gardeners soak their peas in paraffin to deter mice from digging them up and eating them. Honey is taken from bee hives and one mouse was found walled up in wax in a bee hive, presumably stung and preserved when caught in the act! Individuals will climb trees and bushes to obtain shoots at the ends of branches and the tail is used to good effect to help with balance. When berries, hips and haws are taken, wood mice tend to discard the flesh of the fruit and eat the endosperm in the seed.

Breeding

The breeding season, as determined by female reproductive activity, starts in late February, March or April and ends in October or November. Males mature in advance of females so that they are fertile before the females become receptive. The length of the breeding season may be increased by good food supplies, such as a heavy acorn crop, so that pregnancies occur into and sometimes throughout the winter. Gestation is 25–26 days with a mean litter size of 5.5 in the normal breeding season. The resorption of embryos is occasionally seen, and the sex ratio in the laboratory is 1:1 at birth.

Development from birth is variable between individuals and is often faster in males than females. Young are born with naked pink bodies (about 1–2 grams) and by the third day they show slight pigmentation between the closed eyes, behind the ears and down the centre of the back. By the fifth or sixth day the upper surface is completely pigmented and by the end of the first week fine grey-brown fur is emerging. The back is covered with hair by the ninth day and soon afterwards the pink belly is covered with grey-white fur; at this time the tail and feet darken above so completing the greyish brown juvenile coat. The eyes open at about 12–16 days and weaning occurs at 18–22 days from birth. The weight at weaning is about 6–8 grams. Growth in summer is rapid so that individuals are likely to reach 11 grams by six weeks of age and 20 grams by 14 weeks, but there is much variation. Autumn born individuals may take 20 weeks or more to reach 20 grams.

Females may become pregnant when they weigh 12 grams and they regularly conceive again soon after giving birth, so that pregnancies follow one another through the breeding season. Males may become fertile when they weigh 15 grams in summer, but young of both sexes born later in the year remain immature until the following spring, unless food supplies permit winter breeding.

Population structure

Wood mouse populations may reach 100 or more animals per hectare in woodland at the end of the breeding season, but only about 17 per hectare in arable land including ploughed fields. Minimum densities in both habitats reach very low levels so that mice disappear completely from some areas after severe winter weather or food shortage. There was a disappearance, following the cold winter of 1981–2, in two of fourteen 0.81 hectare woodland study areas surveyed for the Mammal Society, and only between one and five mice were recorded on seven of the other study areas.

Numbers may increase from one winter to the next after winter breeding, but there is no evidence of a rise and fall over a 3–4 year period as seen in some vole species. 'Plague' numbers are reported occasionally from central Europe, but after high winter numbers the population may actually decline by the next winter with only a small, short-lived, increase due to breeding in the summer.

The age structure of wood mouse populations, as indicated by weights of individual mice, varies greatly throughout the year and is related to the length of the breeding season. The winter structure, starting at the end of the breeding season, is dominated by late-born young and some older individuals in their first winter, but very few, if any, mice survive their second winter. This structure continues through the spring until the first young of the year, having been born in April and May, join the 'above ground' population in May or June. At this time two groups – 'overwintered animals' and 'young of the year' – can be distinguished by body weight and fur colour (see

Fig 9. After surviving the winter, wood mice die during the following spring and summer. None of this group have lived to see a second winter.

Description). If winter breeding took place there is little or no sign of this grouping.

Most wood mice die young and their mean expectation of life is about 8–14 weeks from weaning during the summer months (although the total life-span in captivity is 18–20 months, or longer). As a result, juveniles do not greatly outnumber the overwintered adults until autumn. This poor survival of juveniles is probably the result of aggression from adult males, and possibly the females, which are intolerant of strange individuals. As the breeding season progresses, most of the overwintered adults die off. Then, as food supplies increase in the autumn, a more tolerant adult population, mostly born during the current breeding season, survives well and the population increases. In woodlands next to arable land, there is likely to be a further addition to numbers in the autumn as mice move from the fields after harvesting. This is followed by a drop in numbers in the spring when the mice move back into the fields.

The net result of these survival patterns is that numbers increase in the autumn and early winter and decline into the following spring, and then often remain relatively stable until early in the next autumn. The extent of the decline at the end of the winter, and therefore the general level of the summer population, is often correlated with the previous autumn's crop of mast (tree seeds) in deciduous woodland. Poor food supplies lead to poor survival over winter, but a good crop usually allows a large proportion of the winter population to survive to the next summer.

In woodland near Oxford, it has been found that a large summer population delays the increase in numbers in autumn by keeping down the survival of juveniles. Thus, there is a density-dependent control on the population which results in winter numbers remaining relatively constant from one year to the next despite the great variations in summer numbers. A good food supply in

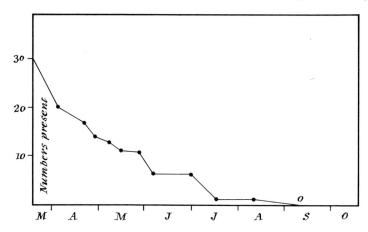

Fig 10. *Typical fluctuation in the density of wood mice in woodland.*

autumn raises the population surviving through the winter and into the following spring and so, in turn, keeps juvenile survival low.

The way that juvenile survival is affected by the adults has been demonstrated experimentally by removing adult males from the summer population. The autumn increase in juvenile survival comes earlier because it seems that the adult males, and possibly the females, keep the summer numbers relatively stable by aggression towards the juveniles. The aggression of adult males is known to increase at the start of the breeding season and this, combined with less food being available, contributes to the decline in numbers of adults in the spring and the poor survival of juveniles in summer. The part played by food supply and predation in controlling population size is discussed below.

Predators and mortality

Bird predators of wood mice include the heron, kestrel, tawny owl, barn owl, short-eared owl, long-eared owl and little owl, and other predators include weasel, stoat, fox, badger, pine marten, domestic cat, grass snake and adder. These are mainly night-hunting animals. Kestrels take wood mice turned up by the plough during the day and they sometimes hunt into twilight when the mice are active.

The wood mouse is known to be particularly important in the diet of the long-eared owl in Ireland where it makes up about 70 per cent of the prey. This compares with only 25 to 50 per cent in Britain where the short-tailed vole is also available. Similarly, the wood mouse makes up 75 per cent of the barn owl's diet in Ireland compared with 11 per cent in Britain. Wood mice are the

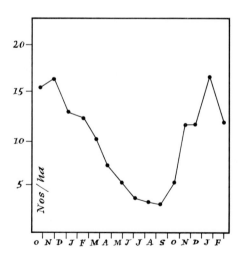

favoured prey of tawny owls in spring, before the plant cover on the woodland floor has grown. In an English deciduous woodland tawny owls made up 30 per cent of their diet with wood mice, and predation by tawny owls seems to take a greater proportion of the wood mouse population when it is low than when it is high.

The impact of owls and other predators is difficult to assess. In Scandinavia, in mixed grassland and woodland with a large variety of predators present, the annual births of wood mice match the estimated number taken by the predators. In Britain the impact of predators on the wood mouse population may be seasonally important, especially at low mouse densities, but more precise information is needed.

Dispersal may account for many of the losses of wood mice observed in population studies but its importance compared with the various causes of mortality is uncertain. Juveniles do not seem to disperse from where they were born but they may die before they are trapped, or soon afterwards so any movements are missed. Adults certainly move long distances in spring and summer, and in woodlands with adjoining fields a large proportion of the spring 'mortality' revealed by trapping may be due merely to mice moving out of the study area!

Parasites and disease

Many ectoparasites infest the wood mouse, including a number of mites, three species of lice and two ticks – the sheep tick, *Ixodes ricinus*, which occasionally bites man, and *Ixodes trianguliceps*, a common tick of small mammals. Twenty-two species of flea have been found in association with the wood mouse in the British Isles of which seven are commonly recorded. These are *Hystrichopsylla talpae talpae*, for which the wood mouse may well be the true host, *Typhloceras poppei*, which is very closely associated with the wood mouse but not throughout the host's range, *Rhadinopsylla (Actenophthalmus) pentacantha*, which is frequently found in the nest as well as on the mouse. *Ctenopthalmus nobilis nobilis*, *Ctenopthalmus nobilis vulgaris* and *Malaraeus (Amalaraeus) penicilliger mustelae* all of which are common and *Megbothris turbidus* which is found frequently but not throughout the host's range. The small beetle, *Leptinus testaceus*, is sometimes found on mice and in their nests, but it is not a true parasite.

Wood mice may carry ringworm fungi, they suffer from a tuberculosis condition and in Britain, but not in Ireland, they are commonly infected with the spirochaete bacterium *Leptospira*, which, in some forms, causes Weil's disease. Infection with *Leptospira* increases with the age of the mice and with the water content of the soil. The sheep viral disease, louping ill, has also been recorded from wood mice. A wide range of internal parasites, including protozoa, tapeworms, flukes, and roundworms have been found in wood mice.

Relations with Man

Wood mice cause little economic damage except as pests of newly sown sugar beet. If much of the drilled seed fails to germinate, mice are often blamed and evidence of the neat regular holes through which the seeds have been taken, and the remains of the seed coat and clay pellet which covers the seed, are usually enough to confirm the farmer's suspicions. Occasionally mice may be killed by using rodenticides such as 'warfarin' but prevention is better than cure and three factors help to limit the damage.

Sowing is carried out in the spring and the later drilling dates in March and April are less likely to suffer damage as numbers of mice will be lower than in late winter. Also, the seed will be at risk for a shorter period when temperatures are higher and germination is quicker. Wet soil conditions after sowing help to prevent soil from being blown away and exposing the seed, a problem which might occur if the soil was dry. In addition, if the seed bed is too cloddy or the drill is not set deep enough then some of the pellets will only be partially covered, or not covered at all, and the new food source will be more easily discovered. Thus the damage to the drilled seed may be avoided to a greater or lesser extent by careful attention to these points. However, the amount of damage to sugar beet varies greatly from one year to the next and is mainly related to the density of mice in the overwintering populations.

Other pest problems with wood mice include the digging up of newly sown cereal and the digging up of peas, but this damage is not usually serious. In woodland, many potentially germinating seeds must be eaten but often the fruit or seed production is so great that plenty survive and the stores made by mice, if forgotten, contribute to the seedling population by protecting the seeds from other predators. In gardens, bulbs and flower heads may be eaten and wood mice may attack any stored food they can get at.

Wood mice are easily kept in captivity if given laboratory mouse food or pet rodent seed mixtures. They become much less nervous than wild specimens but never completely tame, and they have a nasty bite. Handling is best carried out by holding the scruff of the neck behind the ears. If the tail is held, the base must be gripped firmly because the skin sloughs off very easily and attempts to hold the mouse further down the tail will result in loss of the skin and eventually the exposed tail bones.

Legal status

No legal protection is afforded to wood mice owing to their presence in most habitats and their occasional pest status. However, the Wildlife and Countryside Act (1981) does affect the taking or killing of shrews and this can often be done inadvertently when mice are being trapped. If live traps are set to catch any species of small mammal and might catch shrews as well, whether it is intended or not to catch the shrews, it is an offence not to take reasonable

Typical postures of wood mice in aggressive encounters (after Gurnell).
Left, submissive upright, and right, aggressive upright.
Below, boxing.

precautions to keep the shrews alive. Such precautions might be frequent visits to the traps to release the shrews before they starve to death, or the provision of food such as fisherman's 'casters' (blowfly puparia) or fish- and cereal-based cat food, or the provision of a shrew escape hole in the trap nestbox measuring 13 millimetres in diameter. Strictly speaking, it seems as if a licence is not necessary if the intention of the trapping is only to catch mice and voles, but it is advisable to apply to the NCC for a licence whatever the intention. If a licence is not held, then the precautions to keep shrews alive must be carried out, and the traps must be set for the purposes of public health, agriculture, forestry, or nature conservation: otherwise an offence is committed.

The yellow-necked mouse

The yellow-necked mouse, *Apodemus flavicollis*, is a very close relative of the wood mouse and is easily confused with it. They interbreed on the continent of Europe, but not in the British Isles. Apart from being larger in size than mainland wood mice, on average 1½ times the weight, the yellow-necked mouse only differs externally in having a bright yellow chest spot joining the dark upper fur on either side of the neck – hence the name – and sometimes extending down the chest. There is often a slight difference in the shade of the grey belly fur, being very pale in the yellow-necked mouse and only pale in the wood mouse but this distinction is not always easy to make. Skeletal characteristics, teeth in particular, help to distinguish the species (see pages 4–5), but otherwise there are no clear-cut differences to aid identification.

When held in the hand the yellow-necked mouse is often more vigorous than the wood mouse in its struggling and it is more inclined to bite. Very often it will make loud squeaks which sound like screams, but wood mice will also do this on occasion. Another pointer to identification is the relatively thicker base to the tail especially in the male yellow-necked mouse, associated with the development of a scent gland (see page 14). Body weights range to over 40 grams for adults and head and body lengths are commonly 130 millimetres or more.

The distribution of the yellow-necked mouse is restricted in Britain mainly to the south and south-east of England and to the Welsh border country, with only sporadic records coming from south-west England and the more northerly counties of England. This is difficult to explain, but it may be associated with the distribution of drier habitats and mature deciduous woodland. On a local basis, yellow-necked mice are patchily distributed and are more often found in open areas of woodland than in thick undergrowth. The habitats where they may be found also include coniferous woodland, hedgerows and rarely, open fields in arable land.

The home range is larger than that of the wood mouse and it is believed that runways above the ground in trees and shrubs are used to a greater extent than by wood mice. Although individuals of both species may be found, or rather trapped, in association with each other, it seems that they are generally separated in space by the smaller wood mouse avoiding the yellow-necked mouse. In one area of woodland in Gloucestershire yellow-necked mice were commonly more abundant than wood mice for a number of years, but in Essex woodlands records of the two species showed between one quarter and one fifth to be yellow-necked mice. The patchy nature of their distribution is also shown within one locality where it is common to find individuals in one month, often in winter, but not in the summer.

The population fluctuations of yellow-necked mice generally follow those of the wood mouse except that they are often at a low density and interrupted by periods of absence. The increase in numbers may start earlier in the autumn than that of the wood mouse and declines may take place at a slower rate.

The signs and behaviour of the two species are very similar although the

Yellow-necked mouse in apple store.

larger species probably spends more time in trees. The food, predators and parasites are all presumed to be similar in the two species but the breeding season may be slightly shorter in the yellow-necked mouse; the embryo number in Europe is slightly smaller and there are fewer litters per female. The yellow-necked mouse is renowned for entering houses and outbuildings but is less likely to cause damage to agricultural or garden crops than the wood mouse because of its more limited distribution and preference for woodlands.

Left, defensive posture and nosing.
Centre, upright defensive.
Right, jump attack.

Opposite
1. *Yellow-necked mouse perched in Forsythia.*
2. *Neck patterns of yellow-necked and wood mouse.*
3. *Wood mouse using its tail to balance.*
4. *Wood mouse with ticks.*
5. *Feeding platform in disused birds' nest.*